WORLD
HARVESTERS

WORLD HARVESTERS

BILL HUXLEY

FARMING
PRESS

A catalogue record for this book is available from the British Library.

ISBN 0 85236 302 8

**Published by Farming Press Books
Wharfedale Road, Ipswich IP1 4LG
United Kingdom**

*Distributed in North America
by Diamond Farm Enterprises
Box 537, Alexandria Bay, NY 13607, USA*

*Cover design by Andrew Thistlethwaite
Typeset by Winsor Clarke, Ipswich
Printed and bound in Great Britain by
Butler & Tanner Ltd, Frome and London*

Front cover

Six 979 FMC pea pickers at work for the Owatanna Canning Co. of Minnesota during the 1991 harvest.

Back cover

Norfolk Lavender Ltd of Heacham, Norfolk, has been an innovative company since its founding by Linn Chilvers and Francis 'Ginger' Dusgate in 1932. Its business-like approach has been typified by the desire to mechanise harvest operations. A motorcycle combination equipped with rotary cutter, which provided their first system, was followed by other adaptations until the present machine, built by a local company, evolved in 1970. The sole UK lavender producer with a worldwide market, Norfolk Lavender grows almost 40ha laid out in 'hedge' style and harvested at the rate of 1.25ha per day, usually commencing mid July.

Windrower built by McDon Industries of Canada. These machines are often worked in tandem on larger acreages.

Frontispiece

A Claas flax harvester: state-of-the-art equipment for dealing with an ancient crop that has traditionally been difficult to process.

Introduction

Following the interest created by my earlier book on combines, I felt that further coverage of the fascinating subject of harvesting equipment would be justified.

Rather than confine the contents to cereal and related harvesters, I felt that the vast range of equipment for a large number of crops would broaden the scope of the book and provide additional information and illustrations for the reader. The development of the various types of powered harvesters throughout the world is presented from the earliest up to the latest high-tech offerings in operation today.

Thanks are also conveyed to the numerous manufacturers, dealers, museums, agricultural associations, PR companies and individuals whose names appear in the acknowledgements or throughout the pages.

I should also like to thank Keith Stevens, farming correspondent and broadcaster, for his support and advice, and finally my wife for her longstanding tolerance of my ongoing research that occupies almost all of my leisure time.

I hope that readers of this publication will find it of interest. Queries and comments will be welcome at the following address:

Bill Huxley
46 Loomer Road
Chesterton
Newcastle
Staffs ST5 7LB
United Kingdom
Tel/Fax 01782 566043

Dedicated to the late
Charles L. Cawood,
whose remarkable knowledge
of agricultural tractors and farm machinery
will be greatly missed,
as will his blunt 'Sherman tank' sense of humour
which I personally relished!

Combine Harvesters

The combine harvester brought about quite a dramatic change in farming, especially in reducing the amount of hand labour on the farm, but also in changing the organisation of farm work. Threshing was no longer a winter job. Grain harvest became more dependent on weather conditions than ever before. There was the problem of a crop that must be harvested when evenly ripe and dry, and weeds could cause some extra trouble — a situation that was not as much the case when using the binder and stationary thresher. The combine introduced the need for dryers in case of adverse threshing conditions. Changing over to combines made a complete chain of equipment redundant: the binder, thresher, stationary baler and many times also a stationary engine or tractor to power the machine. The combine represented a huge investment for a piece of machinery that can be used only a few days a year, depending, of course, on the variety of crops and local climate. There was the possibility of spreading machine hours over a longer time by means of contract or custom combining. The growing lack of labour after World War II was an important reason for switching over to combines. The number of people required to fork sheaves and so on simply was not available anymore at a suitable labour cost.

Today the market for combines has become a true replacement market. In some cases combines are replaced because the annual costs for repair and maintenance exceed the cost of a new machine. The relatively limited use of the equipment has left many older but technically good machines still being used on smaller farms. Many combines have been replaced because new machines have a larger capacity which reduces the number of machines and operators required. Better operator comfort from new cabs and controls is often a reason to change a combine. In the past twenty years the yields of wheat on many farms have doubled, and many old combines that are still in fine shape cannot make it to the end of the field any more without unloading.

There were combines in the western United States as early as the end of 1880s, many of them pull-types but also self-

Continued on page 8

The Reverend Patrick Bell, a Scottish minister with an active mind for agricultural mechanisation, is now generally credited with being the first producer of a successful cereal-reaping machine, although there were several other claims to this distinction. His machine (illustrated) preceded the McCormick reaper by a scant three years.

Cyrus Hall McCormick's reaper brought out in the United States in 1831 is usually referred to as the original machine of its type. At about the same time a rival machine was announced by Obed Hussey, ex-sailor turned inventor, whose product closely rivalled McCormick's machine for several years. Although Hussey's patent was in force prior to McCormick's, his first reaper did not operate until 1833.

propelled machines. It was after World War II before combine use became established in Europe. The first combine in the Netherlands was a pull-type Caterpillar-Holt in 1934, which was tested on a land reclamation project. The machine worked all right but had difficulty in handling the amount of straw produced on this kind of fertile soil. Claas of Germany sold its first combine in 1936 and had made some 1400 units by 1941. Massey Harris was the first self-propelled combine on the European market in 1945, and it very soon found a number of buyers. In the Netherlands there was one combine in 1945, there were five in 1946 and about 250 in 1947. By 1955 there were over 2000 combines in Europe and the stationary thresher was about to disappear.

Many parts of Europe found that American combines had some difficulty in dealing with the large amounts of straw that grow in wet climates. Capability of handling amounts of tough straw has always been a major consideration in European combine design. Partially for this same reason, the walkerless combines with their 'rotary principles' have not been as widely accepted in Europe as in other parts of the world. Straw is still an important product on many European farms. In the Netherlands, the majority of the straw is baled

Events such as the Great Exhibition held at the Crystal Palace in London in 1851 were a virtual shop window for the new reaping machines then appearing. While McCormick's machine was certainly not the first reaper to be built, his determination in developing and marketing a successful machine stood him in good stead, and English companies were soon licensed to manufacture these highly acclaimed labour-saving machines from America. One company that did not adopt the licensing arrangement was the Samuelson Co Ltd. Its founder, Bernhard Samuelson MP, was the first president of the Agricultural Engineers Association in 1875, and his name is carried by the current headquarters of this association. The machine shown is one of Samuelson's earliest self-raking or sail reapers.

and marketed for animal bedding but also widely used by the flower bulb growers. Until the early 1970s, straw was also an important product for the cardboard industry. In the northern part of the Netherlands, where cardboard was made out of straw, the combine was not as quickly accepted as in other parts of the country., The cardboard industry was based upon the larger sizes of straw bales that came from stationary presses and simply refused to handle the smaller-sized bales produced by the pick-up baler. Despite the fact that separating kernels from an ear is pretty much the same process all over the world, there are always a number of reasons why things develop in different ways at different places in the world. Contrary to other harvesters, the combine market has developed almost completely into a self-propelled market, rather than pull-type.

Hydraulics and electronics are now commonly used to improve combine efficiency but have not changed any of the basic principles. The demand for greater capacity continues, but has not led to dramatic changes of the concept. It has only reduced the number of new machines the market can absorb. The combine without doubt is well-established around the world, still reducing the agricultural workforce.

MARTIN SMITS

A	B	C	D	E & F	G & H	I
Spout where the Corn comes out when the Machine is used as a Single Blast Machine.	Spout at which Chobs are delivered.	Second Dressing Apparatus.	Feeder.	Untying Sheaves and handing them to the Feeder D.	Men on Stack pitching Sheaves to E and F.	Engine Driver.

Ransomes & Sims' portable steam engine (8 nominal horsepower) and stationary thresher are shown as depicted in their catalogue of 1860, with explanatory data which suggests that this system was quite new at the time. Later, under their full name of Ransomes, Sims & Jefferies Ltd, they were to become one of the largest builders of agricultural machinery in the United Kingdom, with a range that included traction engines, threshers, balers, ploughs, combines, tillage equipment, grass machinery, etc.

9

A typical combine of the mid to late 1800s is illustrated. By the 1880s the Best and Holt companies were among the most prominent manufacturers, and production continued for a number of years after their 1925 merger which created the Caterpillar Tractor Company. For hauling the heavy, cumbersome landwheel drive early models, upwards of 40 or more horses or mules were required which, if spooked by fire or component failure, would take off with dire consequences for the machine and its operators! Independent steam engines mounted on the harvesters to handle the cutting and threshing operations reduced the number of animals required, and the arrival of steam threshing engines further reduced the dependency on animal power. With the advent of the gasoline engine and tractors their use was totally eliminated, albeit slowly. Sacramento Valley farmer G.S. Berry notched up quite a few firsts in the 1880s with a self-propelled steam model, a straw burner and a 40ft header which, although very large for the period, was on occasion exceeded in size.

As with the early development of threshing machines, the initial 'combined harvester' designs were numerous and varied in their capabilities and practicality. Moore & Hascall's widely acclaimed Michigan combine of c.1836 was generally credited as the first machine of note. This machine was used in the mid 1850s in California where wheatlands well in excess of two million acres became the ideal proving ground for these novel machines.

Attractive early advertising from the Massey Manufacturing Company featuring the Toronto light binder, which won first prize at the Paris International Exposition at Noisel, France in 1889. The ad is clearly indicative of the firm's appreciation of publicity and good marketing on a worldwide basis.

A typical early German harvest scene. Heinrich Lanz of Mannheim was a prolific builder of steam engines, threshing machines and other farm machinery and eventually became renowned for their range of single-cylinder Bulldog tractors on which the British Field Marshall series was based. The Bulldog tractors grew popular for hauling threshing machines and baler sets and for performing the heavier tasks on large arable farms. Shown here are a Lanz steam engine and thresher.

During the first decades of the 20th century, drivers were, of necessity, considerably adept at handling the huge teams of animals that were required to pull the enormous machines of the time. So much so that contests were often included in such events as Fourth of July celebrations, etc. A useful aid in handling these teams, which was developed by the Schandoney family, involved the Schandoney hitch, shown here, that required the control of only the few lead animals. The most noted historian of this period was probably the late Al F. Higgins, whose archive collection is now held by the University of California in Davis.

Because of its considerable size and favourable climate, Australia figured early in harvester development. Following on from earlier stripper development undertaken by Ridley, Bull and others from the 1840s, the Hugh Victor McKay Harvester Company became one of the best-known names down under. Established in 1886, their Sunshine brand name was soon known worldwide and their first self-propelled stripper harvester appeared in 1909. They were eventually absorbed into the Massey Harris organisation. Illustrated in operation during the 1926 harvest is a Sunshine McKay header harvester hauled by a Hart-Parr tractor.

European cereal growers had to wait much longer for the arrival of the combine harvester than their North American and southern hemisphere counterparts, the general reason given being climate and the smaller fields which did not suit the cumbersome machines of the time. The first production machine in the UK and possibly Europe was the Clayton and Shuttleworth of 1928, which is illustrated. This 'Combined Harvester and Thresher' was shown at Rushall Manor Farm, Pewsey, Wiltshire by the Institute of Agricultural Engineering, which had organised the demonstration for evaluation purposes.

Contrary to popular belief, the stripper-header system of harvesting, whereby only the crop was recovered, leaving the straw intact, is a very old concept. It was initially described by Pliny, the noted scribe of the Roman period, and involved a wooden machine that was pushed through the crops by oxen and operated in the fashion of a comb.

Joe L. Thomas has long been active in both agricultural and commercial preservation and research. He has spent a lot of time down under and his photographs shown on these two pages feature (left) a 12ft cut Sunshine harvester of 1928 powered by a Fordson skid unit, which could achieve a bag a minute, and (right) a 1920s pull-type 6ft cut header harvester by John and David Shearer.

16

Facing page

In Europe, combine harvester development began in earnest during the late 1920s and early 30s when the Gebrüder Claas Company of Westphalia, Germany, headed by August Claas, became alerted to the possibilities presented with the importing of a batch of American combines. They began development work on a 'wrap-around' design that was possibly influenced by the earlier Baldwin/Gleaner tractor-mounted machines. This initial undertaking was followed by a conventional pull-type design which could have been the result of a visit to the United States by agricultural experts who viewed, amongst other machines, the 'high speed' harvester then under development that in 1935 became the celebrated, trail-blazing All-Crop 60. This machine, known as the Mähe-Dresch Binder, was announced in 1937 and was the forerunner of the highly successful Super series of the 1940s. It is shown to the left.

As this book deals with powered machinery, only a brief mention will be made of horse-drawn harvesters. Shown to the right are a peanut picker (above) and an engine-driven potato harvester (below), both from the Massey Harris company. Beet and potato lifters/ploughs, being of a basic construction, were manufactured in large numbers by a host of engineers, blacksmiths and assorted companies in most countries of the world.

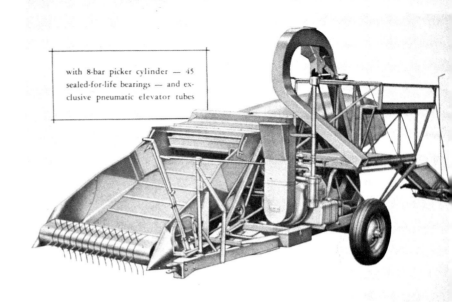

with 8-bar picker cylinder — 45 sealed-for-life bearings — and exclusive pneumatic elevator tubes

In 1931 when Allis-Chalmers acquired the Advance Rumely company in La Porte, Indiana, they inherited a range of threshers and combines that included hillside models as required in the north-western region of the United States. The levelling device on these machines consisted of a hinged shaft-driven header and the left-hand wheel mounted in a sturdy swinging frame which was worm-gear driven by the combine engine and activated from the tractor seat. However, with the rights and the basic designs of the All-Crop 60 soon to be acquired, for which H.C. Merritt ordered total commitment in terms of modification and updating, the heavy, bulky Rumely line was gradually phased out. A No. 3 machine was imported into the UK in 1934, and following the replacement of the standard 16ft header with one of 12ft, it handled our heavier crops quite well. A similar machine is illustrated.

The first Allis-Chalmers forage harvester of 1940 (illustrated) was based largely on componentry from the All-Crop 40 harvester which, when introduced in 1938, saw the adoption of automotive-style assembly line technology and materials that greatly increased production. When it was reintroduced in 1948 after the war, several modifications had been made, one of which was the use of 'cut and throw' curved knives. This model, however, suffered from a lack of development for UK farming operations and consequently never took off in the market.

The Fahr company of Gottmadingen, Germany commenced combine development in 1939. Their first machine (illustrated) was a tricycle-type self-propelled design but of a pull-type layout with the header on the right side. As the owners of both the Fahr and Ködel & Böhm companies, Klöckner-Humboldt-Deutz integrated combine production at the Lauingen plant from 1970.

The first commercially successful self-propelled combine from Massey Harris was the No. 21, derived from the No. 20 which the legendary Tom Carroll had received the brief to develop. Of a size somewhat larger than usual, it was the last model from MH to be classified as a 'Reaper Thresher'. After this time they referred to their machines by the current name. Without doubt the most notable success of the 21 was the 500 machine 'Harvest Brigade' which travelled 1500 miles up through the United States during 1944, achieving 1,019,500 acres harvested, over 25,000,000 bushels of grain recovered and a third of a million man-hours saved as against the use of pull-type machines, a clear indication of Tom Carroll's expertise and influence.

The example shown to the left was being demonstrated at Craige Farm, Leuchars, Scotland for Andrew Small by dealers J.B.W. Smith of Cupar, Scotland during the late 1940s.

Originally developed in the late 1930s but not introduced until 1946, the Claas Super (left) and its variants were highly successful machines that remained in production for 15 years. For power take-off operation, combines of more than 35 hp were required, but when combines were equipped with an auxiliary engine, 20 hp was sufficient for propulsion only, and this would have been within the capabilities of the Fendt tool carrier and similar tractors. Another useful feature of the Super was its airplane wheels and tyres which, being of the balloon pattern, made for lower rolling resistance and reduced soil compaction.

The once ubiquitous Allis-Chalmers All-Crop 60 combine, which was developed from an original 1930 design by Bob Fleming, set the standard for compact combine harvesters from its introduction in 1935. This is borne out by the fact that its production life exceeded 20 years. Available with tanker or bagger option and pto or engine driven, it proved successful in harvesting over 100 types of crops, thereby fully living up to its name.

Facing page

Facing page
Massey Harris were quick to enter into production with self-propelled machines following the considerable amount of publicity generated by the 'Harvest Brigade' operation (see page 20). In 1946 their pull-type Clipper, designed as a compact straight-through machine to compete with the A-C All-Crop 60, was adapted to the self-propelled mode, and the No. 726 appeared in the UK as the first European self-propelled machine in 1949.

It was produced at the Kilmarnock Scottish plant as the model 726 — the 7 prefix denoting UK origin. Although shortages continued in the immediate post-war years, manufacturing techniques were boosted by the adoption of aero industry technology such as the use of unified threads, hydraulics, assembly line methods, etc. In the photograph shown, a trio of 726s, a notable feature of which was the electric motor table lift, are awaiting delivery outside the J.B.W. Smith dealership in Cupar, Scotland

The early 1950s was not a good time for small companies to release new machinery: a dollar crisis added to the reluctance to accept radical designs could well have been the reason why the M.B. Wild 'Harvester Thresher' shown above, whose only working parts were blower, awner and rotor, failed to achieve any notable commercial success.

The early Allis 'B' would have had no problem with this lightweight, independently powered machine.

Harvesting durum wheat in South Dakota during 1951 with three John Deere model 55s and two experimental Allis-Chalmers model 100s. The machines shown were operated by the Wertz brothers. Although initiated in 1947, the All-Crop 100 did not become commercially available until 1953 and was superseded within two years when Chalmers acquired the Baldwin Harvester Corporation.

Just about discernible is the outline of the Allis-Chalmers All-Crop 40 harvester in this self-propelled adaptation by the National Institute of Agricultural Engineering in the 1950s. Undertaken at the behest of the National Institute of Agricultural Botany, it was followed by four further machines by the E. Allman Company. Originally designed as an inexpensive pull-type for the 50-acre farmer, it introduced automotive-style assembly-line production to the agricultural machinery industry.

Lanz arrived on the self-propelled combine scene in 1954 with the machine illustrated, their MD240S, self-propelled model of 2.40 m header width and powered by a 55PS Opel Kapitan motor. The year 1956 was significant for this old-established company, as news of the two-hundred-thousandth Bulldog tractor was closely followed by the new firm name of John Deere Lanz. By 1960 full integration into the Deere organisation resulted in the end of the Lanz name.

Facing page
When announced in 1953, the SF was the first self-propelled model from Claas of Harsewinkel. Advertised as the combine with the hydraulic variable drum speed, it offered header sizes from 8½ to 12ft and a straw press and grain bin option. It was produced until 1961, when superseded by the Matador series.

Harry Ferguson established manufacturing facilities with the Standard Motor Company of Coventry in England following the termination of his agreement with Henry Ford, who had built tractors incorporating Ferguson's revolutionary hydraulics system during the early 1940s. Commencing in 1946, the TE20 tractor (illustrated below) and its variants proved an instant success with the revolutionary hydraulics for which a full range of implements soon became available. One of these was a 'wrap-around' combine developed in the early 1950s, but because of the impending merger with Massey Harris, who were developing a compact self-propelled machine which would be known as the 735, further development on the Ferguson machine was shelved.

Directly or indirectly, Allis-Chalmers farm machinery covered almost the whole spectrum of agricultural mechanisation. Illustrated is a Hurricane Junior sugar cane harvester by the long-established Thompson Company of Thibadeaux, Louisiana, and just showing its face in centre is a B15/B125 power unit more commonly seen on Chalmers' own Roto Baler and All-Crop 60 harvesters, etc. Allis-Chalmers also supplied this company with UC high-clearance tractors equipped with GM two-stroke diesels, HD-6 track-type tractors and others for mounting various sizes and types of sugar cane machinery. Machinery custom-built for A-C included sugar beet and potato harvesters produced by Oppel Inc. (of Oppel wheel fame) during the 1960s.

As a considerable area of Europe consists of small family-size farms, Gebr Claas in 1955 announced their compact Hückepack designed with this type and size of farm in mind. Based on a tool frame tractor, it highlighted the early European inclination to diesel power. As it could readily be replaced by a wide range of attachments, it proved an economical and versatile machine, and was produced until 1962.

Recently retired after giving sterling service is this Allis-Chalmers Gleaner, which was one of the original twenty imported into the UK during 1957 and is still in the hands of its original owner, Les Dessurne of Newark, Nottinghamshire. Les and his family now operate a younger model (1964) Super A built at nearby Essendine, Lincolnshire, which is following in the old-timer's footsteps in turning in an almost faultless performance at harvest time. The last UK designed and produced A-C combines were the 5000 of the late '60s which incorporated 'down front' threshing and proved that some of the so-called new ideas are not so new at all!

Long noted for their stationary and pick-up balers, Jones Balers Ltd (David & Glyn) of Mold, North Wales, entered the combine market in the late 1950s. Shown to the left is their Pilot model and above is a largely restored Cruiser on display at the Greenfields Heritage Centre.

When Jones were acquired by Allis-Chalmers in 1961 their machines were superseded by the more widely known

Gleaner combines. Shortly prior to the A-C take-over the Jones brothers obtained an order for 10,000 of their Star conventional pick-up balers from Allis-Chalmers of Milwaukee, which possibly prompted the ensuing acquisition. The principal Mold plant continued in operation under A-C ownership and in addition to balers and other implements, provided components and other combine back-up for the works at Essendine in Lincolnshire. The English A-C operation was taken over by Bamfords of Uttoxeter Ltd in 1971, who promptly dropped combine production in favour of their existing Volvo distributorship, which itself was soon terminated. Balers continued to be available from Uttoxeter but were of a badge-engineering exercise with Bamford, Jones and A-C machines being offered in their individual liveries. Following the withdrawal of support by a major agricultural dealership, the company went into terminal decline in the 1980s when several changes of ownership failed to maintain a satisfactory market share.

In 1964 the Sperry New Holland Company acquired the Belgian company Leon Claeys, makers of the first European self-propelled combine in 1952. Before this time Fahr of Gottmadingen had bought in the Claeys M80 combine to round out their range, which did not include large machines.

The Deutz-Fahr M660 dates back to the 60s when it was also available in pull-type mode as the M66TS (above). With a production build exceeding 10,000 units, it was continued as the 33.70 and proved very popular for trial plot conversions. The M660 to the right is at the National Institute of Agricultural Botany in Cambridge.

34

Manufactured in the former Eastern Germany, Fortschritt combines, which were introduced in 1963, were based upon Russian designs. By the 1980s their more advanced models had become relatively popular, with the largest available threshing cylinders and crop wall sensors being two of their attributes. Following the fall of the Berlin Wall and the reunification policies that were introduced, the gigantic Kombinat Fortschritt was privatised. The Landmaschinen VEB group commenced and the Mähe-Dresch-Werke name, together with a change of colour, was adopted for machines produced after 1990. A further change of ownership has taken place. The early model E 175 shown is equipped with pick-up reel which helps to lift laid and tangled crops. Swathers and windrowers for this indirect form of harvesting are also produced by this company.

Ködel & Böhm, founded in 1870, built their first thresher in 1890 and went on to become the largest manufacturer of these machines in Europe, producing 12,000 annually. The one-hundred-thousandth to carry the Köla label left the factory in 1936. Combine development commenced in 1940, their first machine appearing in 1951, and hydrostatic machines like the ones shown below in 1966.

When introduced in 1961, the Claas Matador and its variants may have seemed rather conventional. However, thanks to their well-engineered design and construction, they became highly regarded and many remain in use today long after 1969 when production ceased. They are popular with agricultural colleges for instructional purposes, and many small-acreage farmers still find them well suited to their needs.

Because of their long-standing involvement with the production of threshing boxes (machines), Ransomes, Sims & Jefferies naturally took an active interest in combine harvester development, which, according to a long-time employee, began in the late 1940s when they experimented with a cutter bar and conveyor system mounted on a stationary threshing machine which was then hauled around the field in the manner of a combine. The sheer bulk and weight of the machine plus the fact that the basic road wheels would have been unable to hold the machine up in soft-going were some of the reasons that made this a non-starter, but Ransomes soon adopted a more conventional method of combine harvester production by negotiating with the Swedish Bolinders-Munktell company for the manufacture of their pull-type machines in 1952. From 1953, in keeping with their agreement with the Ford Motor Company to

supply matching equipment, they were for a time badged as Ford-Ransomes (F-R). Ransomes went over to building self-propelled models, but even though their later models such as the Crusader illustrated above had built up a good reputation, production ended in the early 70s for commercial reasons. Bolinders-Munktell themselves were absorbed by the Volvo company, whose own combine production also finished at about the same time.

Vicon of the Netherlands were one of several manufacturers who opted for a compromise design rather than offer either self-propelled or pull-type root crop harvesters. This option consisted of mounting a tractor skid unit to power the harvester and to provide propulsion. The advantage of this was that, after the finish of harvest, the unit could be removed, the wheels replaced and the farmer then had an extra tractor. However, the trouble of mounting and dismounting the unit plus the fact that farmers usually required additional tractors at other times than the dead of winter seems to have affected sales. Improvement in conventional designs and increase in tractor horse power also reduced sales. UK farmers tend to prefer pull-types, whereas their European counterparts favour self-propelled units. The late 1960s potato harvester shown above features an MF unit but most proprietary brands of tractor were suitable.

There can have been few, if any, more successful or enduring joint ventures than that of the Franz Grimme company of Damme, Germany and the Richard Pearson company of Frieston, Boston, Lincolnshire, which commenced in 1963, when complete harvesting systems were in their infancy, and .

lasted until 1993. By the early 70s they offered a range of potato harvesters to suit all UK requirements. The 1973 Commander shown above, demonstrated by Barry Burrell, lifted 25 cwt every six minutes at 6 mph.

The classic work-hungry look of the Gleaner is well illustrated on the facing page in this M2 equipped with a corn head. When the L and M series were introduced in 1972/3, they heralded several innovative threshing features and introduced electro-hydraulic controls for easier operation. The M, slightly smaller than the L, carried headers up to 22ft and was available as the MH hillside model shown to the right, which could handle slopes of up to 43%. The silver or galvanised finish, popular in the United States, was deemed unacceptable for the European market, and thus all machines leaving the Essendine plant as of February 1958 were finished in the standard 'Persian orange' livery.

Self-propelled harvesters have in recent years had something of a mixed reception in the UK. Since the early 1970s Richard Pearson have supplied several, including the early version based on their well-known Grimme Continental model (left). More recent developments include trials with a self-propelled machine based on the Q Continental principle powered by an air-cooled Deutz diesel, hydrostatic drive, four-wheel steering and with the capability to lift potatoes, onions and carrots.

Facing page
In 1977 Massey Ferguson introduced their unique Power-Flow table (header). This consisted of a 500mm conveyor belt added between the cutter bar and the auger in order to allow head-first delivery of the crop so as to get better threshing and separation results. Prior to this, other MF innovations had included their cascade shaker shoe and the MF Quick-Attach system which enabled the rapid exchange of header systems, such as corn head to cutter bar, etc. MF also introduced fully air-conditioned cabs for combines with their 760 model.

Facing page
The fiftieth anniversary display of the Gleaner at Las Vegas in February 1973: (left to right) 1960, 1951, 1926 and the original wrap-around built on the Fordson model F tractor.

Although somewhat erratic in his business ventures, Curtiss Baldwin was innovative and forward-thinking in terms of harvester development, with the rotary threshing system, the first corn head and his own wrap-around design based on the Savage harvester being some of the pioneering ideas attributed to him and the Gleaner company during his term there. In February 1955 the Allis-Chalmers Company purchased the Gleaner Harvester Corporation of Independence, Missouri.

J. Freudendahl of Denmark, better known as JF, produced over 25,000 of their wrap-around combines between 1961 and 1986, and exported them to more than 40 countries worldwide. The example shown above, attached to an MF 135, was the mainstay of the Northern Ireland Plant Breeders operation during the 60s and 70s until replaced by a custom-built Hege 125 compact machine.

This is something of a mystery machine photographed by a member of the Northumberland Vintage Tractor Club while on holiday in Eastern Europe. It is believed to be a double rotor thresher designed to cater for the less than ideal conditions in some regions in that part of Europe and the Balkans.

Another view of the All-Crop 60, also known as the Corn Belt Combine and the Successor to the Binder — just two of the titles bestowed on this ultra-successful machine for small farmers. As annual production figures exceeded 20,000 units, most other major manufacturers tried to follow suit, but in spite of similar-sized machines from International Harvester, Massey Harris, Minneapolis-Moline and others, its reputation reigned supreme. The restored bagger shown, probably the first on the UK preservation scene, is owned and operated by Brian Steele of Marks Tey, Essex.

The H.P. Bulmer Company, the leading UK cider producer, have an obvious need for mechanisation. The Edwards Vibro harvester (left), which Bulmers had an active involvement in during its development, was derived from long-standing United States mechanised apple and fruit harvesting expertise. Tuthill-Temperley are well-known producers of hoover and other types of fruit harvesters, while the Bulmer organisation have their own extensive in-house engineering department that builds machinery to exact company requirements. Mechanised harvesting has been found to eliminate tree damage that can occur with traditional labour-intensive methods, and pears, plums, damsons, lemons, greengages, oranges and other fruits can be successfully harvested by mechanical means.

Continued overleaf

Above
An International Harvester tractor-based machine for harvesting apples.

Left
A Tuthill-Temperley hoover-type harvester.

48

The Development of Pea Harvesting

Peas have been a popular vegetable for a very long time. However, removing them from their pods has always been an onerous task, requiring several pairs of busy hands for a considerable time to provide sufficient for a family meal. There was no alternative to this until the late 1880s when a French woman by the name of Fauré applied her considerable intellect and designed a mechanical pea sheller or huller, thus opening the crop up to commercial uses.

Her invention consisted of a huge oblong wooden frame into which she mounted a reel (or drum) that was enclosed by string nets and could be rotated by a series of belts, pulleys and handwheels. Within the outer reel a concentrically placed six-sided wooden drum with angle blades rotated in the same direction as the outer reel but at a higher differential speed (beater drum).

Pea pods, hand stripped from the vines in the field, were brought to the viner, as it was called, and fed into the space between the outer and inner drums via a feed chute. They were then carried up to the top of the machine on the longitudinal beams and upon falling off were struck by rapidly rotating beater blades. The hefty impact caused the pods to burst open, releasing the peas which could then escape through the nets (screens).

Below the reel an inclined belt (or apron) which was travelling in an upwards direction and caused to vibrate by means of its square and triangular rollers, carried waste such as leaf, stalk or pod shells over the top and induced the peas to roll down the apron and into a collecting trough. From here they were put into baskets and taken off for processing, i.e. canning.

The entire machine was made from wood and canvas and was hand powered through gear trains, belts and pulleys. Between the years 1890 and 1930 the only significant improvements in the viner were materials, manufacturing techniques and the availability of electrical power but not the basic principle of operations: frames became RSJs instead of wooden blocks; metal covers and panels replaced plywood; rubber and composition replaced canvas for belts, aprons and nets; electric motors replaced a man labouring away at a handwheel.

The situation remained thus until the late 1950s / early 60s and it was quite common to see static viner sites both on farms and at processors' premises with anything between a single viner and a set-up of 64 machines, churning away all day long and under floodlights by night.

The drawback to such a system was the enormous amount of labour required to keep it working. In the field, tractor-mounted cutters cut the vines and laid them into windrows; then green crop loaders came along, picked them up and put them onto tractor-drawn trailers. These trailers transported the vines into the factory, dumped them for hand feeding into the viners and then brought the waste back to the field for spreading preparatory to ploughing in or took it to a location for stacking up into a clamp for high-protein winter animal feed.

Back at the viner site men (and strong ones at that) toiled long hours heaving vines by pitch fork into the viners. Threshed-out peas were put into either 40lb trays or bulk boxes to be fed into the process line.

Some processors even had pea conveyors direct from the viners into their pea-cleaning equipment. This material had of course to be pre-cleaned before it could be processed and had to be passed through air separators to remove light waste, through graders to remove under- and over-sized material and finally through flotation washers which were capable of removing any residual waste material as well as sorting out mechanically damaged peas, splits and such like.

In the late 1950s attempts were made to 'take Mohammed to the mountain' by mounting a static viner in a trailer and powering it with a diesel engine. This unit was towed into the pea field, where it was parked up in a corner and the vines carted to it. This removed the necessity of trailing the vines from field to factory and the waste from factory to field but still involved cutting, windrowing and loading of the vines into the trailer. Vines also still had to be fed into the viner.

By the early 60s the first genuine mobile viners made their appearance on the UK pea scene. They were essentially static viners mounted on their own chassis and towed by a tractor, the big difference being that they were equipped with a pick-up system which made the machines self-feeding. The vines still needed to be cut and windrowed and the tractor went along the row straddling the vines, with the pickup elevator lifting the row and passing it into the thresher chamber. Several manufacturers, primarily FMC and Mather and Platt in the UK, made versions of the mobile viner, all basically similar but different in refinements. Automatic self-levelling was incorporated as well as on-board cleaning with aprons and fans and pod-eliminating systems. The machines became more sophisticated by using hydraulics more extensively.

During the 1960s and early 1970s the mobile viner remained essentially the same in basic concept but with different manufacturers trying different ideas to improve their machines. One significant alteration by Mather & Platt was to lengthen the thresher chamber by some 30%. This had an immediate effect on the capacity of the machine.

However, the greatest single impact on mobile vining was introduced by FMC in 1973 in the form of the multi-beater system. Whereas all mobile viners had previously had a single centre beater and hefty beams in the thresher reel, the FMC innovation had five beaters; one centre beater and four satellite beaters. The longitudinal beams in the threshing reel vanished and were replaced by a smooth continuous inner surface moving at higher speeds. This Planetary Threshing System, as it was called, was still incorporated in a trailed harvester but the entire system indicated that, if input were reduced, it would be capable of much higher speeds over the ground.

FMC next produced an experimental picking reel which was mounted on the front of a tractor and had a system of transfer conveyors to pass the picked materials over the

tractor and into the viner. This worked probably better than was ever imagined because only the pods and some of the lighter leaves and vine tops were picked, leaving the bulk of the vines still rooted in the ground. This greatly reduced uptake of material married in perfectly with the Planetary Threshing System, and by 1975 FMC had produced a complete integrated one-pass, self-propelled Pea Picker (as it became known). The unit was totally self-contained, with its own power pack, a 160hp Deutz air-cooled engine driving all the systems including hydrostatic transmission. It became an immediate success. In addition to reducing manpower and removing tractors and cutters from the field, the picking system itself was far more efficient at gathering the pods from the vines. An improvement in yield of some 7 to 10% was a tremendous bonus.

Of course, once the 'pea picker' door had been opened by FMC, other manufacturers followed suit and introduced their own version of this self-propelled single-pass machine

incorporating their own particular systems, generally of two or three beaters but never five like FMC.

FMC continued development of their model 679, constantly improving it with such things as a bigger engine (213hp instead of 160hp), infeed augers instead of belts, bigger and better buckets, etc. In 1984 FMC introduced a second generation Pea Picker — the model 879 which had 40% longer threshing reel giving anywhere up to 50% greater capacity. One 879 was capable of harvesting over 1000 hectares of peas in a 35 day season. Needless to say, a bigger engine was needed for this larger machine, and a 295hp Deutz air-cooled engine was selected. This is more fuel efficient than its predecessors, using around 30-35 litres of fuel per hour, which is as much as 25% less than an equivalent water-cooled engine.

This new bigger, heavier harvester necessitated a serious look at the problem of compaction and damage to soil structure, and the outcome was the use of ultra-wide

With a time-span of approximately two hours between picking and processing, the operation of pea harvesting has to be performed with almost military-style precision and is a totally continuous undertaking. In the UK, the season usually lasts from early July until six weeks or so later.
The Processors and Growers Research Organisation recently celebrated their fiftieth anniversary, which coincided with the initial mechanisation of the industry. The early method of harvesting involved the use of standard cutter bar mowers, aided by tractor-mounted buck rakes or by towed 'cut lift' machines loading directly into trailers for delivery to strategically located viners. Tractor-mounted cutter swathers were also in use, but by the 1960s the viners had become mobile, thus eliminating wasteful haulage of the crop and surplus haulm. Illustrated on page 50 are a swather (left) and a pull-type viner (right). To the right is an FMC 879 self-propelled machine, negotiating a tricky turn on the move to a new location.

(770mm) Michelin radial tyres. Designed to operate at low inflation pressures, these achieved a reduction in ground pressure of almost 50% over previous types.

A further significant development occurred in 1989 when FMC introduced the model 979 Pea Picker with hydrostatic drive to all six wheels, giving a wheel loading of only 4 tonnes and also providing better capabilities on wet ground than possible with four-wheeled models, as well as reducing soil compaction. Another bonus was achieved with the adoption of a larger threshing drum, which gave an additional 35% capacity. Today, this model is operational in 19 countries ranging from New Zealand up to Norway.

Development of Pea Pickers by all the manufacturers involved continues. It is difficult to forecast what the next 20 years will bring compared with the total revolution in pea harvesting over the previous 20 years.

GORDON BRIDGE

The Mekong Delta in Vietnam includes 2.3 million ha of rice. Quite an expanse, but as 50,000 of these axial-flow threshers have been produced since 1974 by the Agricultural Machinery faculty of the University of Agriculture and Forestry in Ho Chi Minh City, life must be a lot easier nowadays! The threshers are based on original designs by the International Rice Research Institute in Manilla, and it is estimated that 300,000 of them have been built by more than 1000 manufacturers worldwide.

Russian Combines

With an area of about 125 million ha of combinable crops in 1989, Russia (the former USSR) naturally required a considerable number of combines. To harvest what represents 21.6% of the world total of grain there were 850,000 machines, each of which handled an average of 150 ha per season.

While Russia's first horse-drawn machine, developed by Messrs A. Vlasenlco and M. Glumilin in 1868 with a grain-stripping rotor system, was far ahead of its time, it was not until 1930, just as with the European countries, that a serious effort was directed toward the manufacture of combine harvesters.

Large-scale manufacturing facilities were set up in 1930 at the Communar plant in the city of Zaporozhye with the first model bearing the title of Communar and having the following specification: threshing capacity of 2.2 kg per second, header size of 4.4 m, thresher width of 80 cm, spike-toothed cylinder of 61.2 cm, straw and chaff collection capacity of 5 to 8 cu m and an engine of 22 kw.

The year 1932 saw the commissioning of the giant Rostelmach plant to produce the S1 pull-type combine with header size increased to 6.1 m, thresher width to 91.5 cm and engine power up to 29 kw.

Further expansion occurred in 1936 with the reorganisation of the Ukhtomsky plant which, until 1927, had been operated by the International Harvester Company of the United States. Here began the manufacture of the Northern model for the north-west USSR region, which suffered from excess humidity. Features of this machine included rotary straw separators instead of platform conveyor straw walkers and five 120 cm rotors, each with 56 teeth. This system provided better straw separation when working with coarse, long stemmed or wet crops and was also utilised by major European manufacturers in the 1980s.

Following restoration and repair of World War II damage, these plants commenced production of the pull-type S6 and the self-propelled S4 models. This model had a threshing capacity of 2.5 kg per second and a header with dual augers and centrally located canvas conveyor.

The year 1953 saw the successor to the S4 in the form of the SK3, which included several notable improvements over its predecessor. These included a one piece auger, capability to follow ground contours in the longitudinal and transverse planes, V-belt variable speed drive, straw and chaff collector and hydraulic control of the cutter and reel actuation, plus other functions.

Siberia and the Baltic Republics have encountered problems with humidity, long-stem crops, weed infestation and lodging. To deal with this situation the Sibiryak, Kolos and Emissey machines were introduced in the 1960s. These differed from US and European combines in that they featured twin cylinders, either two rasp bars or rasp bar and a spike-toothed cylinder. When working in conditions of 10% plus moisture, they proved more efficient than single-cylinder types.

Further innovations appeared in 1971 with the introduction of the Niva series, which had automatic thresher feed control and electronic grain loss monitors. This early use of monitoring achieved a reduction in grain losses in the order of 5 to 20%, making a worthwhile contribution to profitability considering the vast acres involved. Consideration was given to the recovery of the straw for fodder either by chopping and loading into trailer bins or by leaving in swaths for later pickup. The customary chopping and spreading for ploughing in was a further option.

Recent developments to improve the efficiency in harvesting over 200 million tonnes of grain and up to 100 million tonnes of non-grain residue included the introduction of the Don 1500 produced at the Rostelmach plant from 1986. Unique features consisted of a header equipped with finger beater between the auger and conveyor, which improved performance with lodged crops, increased cylinder diameter (800mm), greater concave wrap angle, 4.1 m five-piece straw walkers and electronic monitoring of field and working unit performance. Capable of harvesting a wide range of crops in the direct harvesting mode or with pickup reel for recovery from the swath, this model is attracting attention in the US, where it is handled by Belarus Machinery Inc.

Finally, this brief review of grain harvesting in the former USSR would not be complete without mention of the early 1980s SK10 family of combines. These axial rotary machines designed for grain and rice harvesting feature a longitudinal rotor replacing the conventional threshing section, and there is a rotating grate type, concave as opposed to the fixed variety. Large, curved pad tracks enable them to operate in the soft, porous black earth regions where wheeled models would create problems.

PROFESSOR NICKOLAI I. KLENIN and DR. PAVEL P. SOROKIN
Moscow Institute of Agricultural Engineers

The SK10 family of combines.

The Don 1500.

Population shifts to urban areas and alternative employment plus escalating labour costs are some of the factors that cause growers of traditionally hand-picked crops to mechanise. In 1983 the Toro & Mityana Tea company of Uganda took their first step in this direction when they purchased tea combines from the late Mitchell Cotts company which were of the half-track design with Busatis double-reciprocating, double-edged plain-knife section cutter bar and 107hp Perkins 6.354 diesel power. Following changes in ownership and policies of the above company, further machines were sourced from the Australian Williams Company and were of the full-track layout with cylinder-type cutter bar and powered by a four-cylinder Cummins of 93hp/70kw.

Illustrated are (facing page) the initial MCTC (Mitchell Cotts tea combine) and (above) a Williams machine which arrived in late 1988. John Kilgour of Silsoe College (Cranfield University) and John Burley in Uganda were actively engaged in mechanised tea harvesting projects.

When you cannot obtain a machine to suit your requirements, an obvious but often difficult option is to have your own built. This is exactly what R.F. & M. Mercer, Lancashire-based farmers, contractors and agricultural engineers, did on behalf of a local customer in the early 1980s. Farmers are often an innovative and usually determined breed, and careful thought and planning led to the Jamaffa two-row brussels sprout harvester illustrated, which is sitting on JCB tracks and powered by a Perkins 4-236 diesel. Although sprouts are one of those difficult vegetables to harvest completely mechanically, with a minimum of labour they can be recovered at a respectable rate of around one ha daily using this machine. Since it commenced work in 1982, two further models featuring Track Marshall running gear have been built.

An Iranian order for 75 machines in 1961 suggests that Bisset of Blairgowrie in Scotland was one of the last builders of binders in the UK. The early 50s semi-mounted power drive example shown is owned by Brian Smith of Shrewsbury, Shropshire and has regular workouts at local events.

Hans-Ulrich Hege of Waldenburg, Germany have thirty years' involvement with combine production for trial plots. This originated with their R & D concerning plant breeding. Quite a number of their original 125 models are still in operation worldwide, as can be seen in this 1970s machine harvesting for the Northern Ireland Plant Breeding Station (NIPBS) in the capable hands of Billy Simpson. The Hege organisation have recently moved into mainstream combine production with the acquisition of the privatised East German MDW (Fortschritt) harvester division.

The next two illustrations are also by courtesy of the NIPBS. In the picture to the left, operator Sam Beattie is monitoring the crop in hand on a Danish-built Haldrup 1500 plot machine equipped with an Epson recorder for computerised weighing.

The machine on the facing page, a Belgian Depoortere two-row flax puller, shown with Billy Simpson again, is something of a rarity in this part of Europe. Flax growing has had a chequered history in the UK, with more interest shown in times of crisis when native raw materials need to be used without recourse to importing. As the full stem has to be harvested, the shorter stemmed linseed flax is more popular because it can be combined.

Facing page
This is another product from the Haldrup company, a trial-plot swath-mower with high clearance for crops such as rape seed. The machine featured is owned and operated by the National Institute of Agricultural Botany, which also use Hege, Claas and a variety of other specialised equipment for their vital operations.

Another high-mobility thresher, shown above, is the Hege 122C, suitable for either small-acreage operators or trial plot work. Pto, electric or engine options and three-point linkage make its 600 kg weight well within the capacity of a wide variety of tractors.

The Royal Show, staged in early July each year at the Stoneleigh Showground by the Royal Agricultural Society of England, is the premier outdoor agricultural and countryside event in the UK. For the 1990 show Claas UK Ltd had a special attraction in the form of a 1937 MDB (Mähe-Dresch-Binder), which was their first commercially successful combine harvester. As the name implies, it was a mixture of binder and thresher equipped with sails and canvases that were the norm prior to the development of augers and conveyors, etc. Capable of operating in mobile or static mode, it offered advanced options such as straw chopper, chaff wagon and bale sledge.

Having identified the potential of this new form of harvesting in Europe, Herr August Claas and Professors Vormfelde and Brenner of the Farm Machinery Institute were instrumental in promoting and developing modern harvesting techniques in Europe during the 1930s and later. The machine shown, which was restored by retired Claas Territorial Manager David Schiell in Scotland, is based at the Scottish Agricultural Museum at the Royal Highland Showground near Edinburgh.

While the materials and technology employed in the manufacture of stripper headers are of the latest available to the industry, the basic principle is one of the oldest known to man. Historians have traced this design back to Roman times, when the vallus, an implement styled like a wooden comb, was pushed through the crop by oxen and removed the grain by a combing action. The Shelbourne Reynolds Engineering (SRE) stripper header attached to a New Holland combine clearly shows the before and after of this method of harvesting and explains the much greater outputs achieved as against the conventional system.

The John Deere Titan II combine was a sturdily built machine endowed with the latest technology for handling a wide variety of crops such as corn and maize that test the design and capabilities of equipment to the utmost. Its working partner here, a Kenworth truck, is another quality American product built by the Paccar Company on the West Coast.

The Crop Tiger is a 1990s addition to the Claas range of combines. Basically a universal machine suitable for operating on small farms in tropical and subtropical regions, it has the capability to harvest rice, cereals, sorghum, soy beans and other crops. Running gear is of the rubber full-length type and is powered by an Izuzu diesel with hydrostatic transmission.

Shown here is Cebeco's 400
container chopper that
replaces the Hesston Field
Queen. The machine is at
work at the Ruinerwold grass-
drying facility in the
Netherlands.

68

With the abolition of straw burning within the European Community, the use of straw chopper/spreaders makes the subsequent tillage operation much easier. This Deutz-Fahr machine, seen on what proved to be a successful demonstration during the 1990 harvest, highlights the need for operator protection, which is an important feature on the present Power Liner, Star Liner and Top Liner models from this company. Deutz-Fahr also feature the prize-winning Master Shift control.

Saudi Arabia, with an annual output of 400,000 tonnes of dates, has addressed the slow, costly and hazardous hand-picking method of gathering this crop. The Agricultural Engineering Department of King Saud University, with involvement from John Kilgour and other experts, placed the date servicing platform shown under evaluation in 1989. Basically it is a mobile access platform with two- or four-wheel drive to handle local conditions, and it incorporates several electrical and hydraulic features that allow the operator safe operating conditions.

Calvin L. Schmidt of Ontario grows a considerable acreage of beans which he harvests with Lilleston or Bob 'Bean King' harvesters. He informs me that most of his output is exported to the UK, but I haven't yet eaten any beans stamped with his name!

As they are a soft-skinned vegetable, tomatoes would, to the average person, be deemed a hand-picked crop. However, the Guaresi company of Pilasti, Italy have successfully developed machines to handle this delicate crop, the model 6-89, capable of a 20tph output, being one of their range. Self-propelled sugar beet harvesters are also built by this company.

In the past two decades the collection and restoration of farm tractors has grown by leaps and bounds. In more recent years combine harvesters and other old farm equipment have also attracted the attention of collectors, and at least one combine is on display nowadays at most major rallies.

As the consequence of a long-standing interest in old farm machinery, the Cambridgeshire-based Jolley family found that in September 1991 their collection was getting a little out of hand with the result that a staggering number of machines were catalogued and lined up for auction as shown in the photographs to the left: 53 tractors, 7 combines (two tanker and bagger Massey Harris 726s and two 735s, an engine-driven Allis-Chalmers All-Crop 60, an International Harvester B64 and a Grain Marshall), numerous root crop harvesters, ploughs and general farming effects. Also shown are a considerable number of implements including several types of early tractors and horse-drawn potato lifters and spinners. This was one of the largest auctions of its type for many years.

As late as 1937 a leading agricultural publication cited the need for complete mechanised harvesters for root crops still being lifted by time-honoured hand-picking methods. Full harvesting systems gained in popularity in the 60s, when Root Harvesters Ltd (now KeyAg) introduced an early form of high-tech innovation with an X-ray sorting facility on their Super Duplex potato harvester shown on the facing page.

73

The Silsoe Research Institute
(formerly AFRC Engineering)
in Bedfordshire have had a
long and varied involvement
in the research and
development of farm
machinery and have been
involved with a large number
of specialised projects. One
such concerned crops that are
either difficult or impossible to
harvest in the full mechanical
sense. The vegetable gantry
shown above is a typical
example.

Facing page
Braud, once noted for combine
harvesters in France, are now a
part of the FiatAgri group and are
undertaking production of grape
harvesters. Several self-propelled
and pull-type models are
available, and as a consequence
of dedicated research and
development, vital aspects such as
the basic shaker system have been
refined to give a pulse rate of 350
per minute, a reduction of up to
50%, which greatly reduces
vegetation damage.

Not an ornate form of gate or portcullis, but a six-row maize header in a vertical stance, probably displayed this way to save space at the twenty-fifth Feria Téchnica Internacional de la Marquinaria Agricola (FIMA) exhibition staged at Zaragoza, Spain in 1991. At this prestigious event all forms of major and ancillary equipment can be seen, from a compact tractor-mounted potato lifter made by the locally based Zaga company and designed for the small-acreage Mediterranean and North African operations through to the largest models of combine harvesters. Technical services can be discussed with the experts in attendance on the multitude of display stands.

There are many regions of the world where combine harvesters are neither physically nor commercially viable. To cater for this sector of the market, a number of manufacturers produce compact, easily transportable threshers which, being less complex in build and operation, are well suited to areas without full-scale repair facilities. In their range of machinery Cicoria of Italy have the ATX 1600 portable thresher shown to the left and also compact combine harvesters, all of which employ the rotary threshing system.

76

Cebeco of Steenwijk, the Netherlands build six-row sugar beet harvesters with a seven-row capability for opening up the field. Propulsion is 4WD, and harvesting power is provided by a 343hp DAF engine. A 14 tonne capacity tank is standard, and the lifters and defoliator are supplied by Kleine of Germany.

The Xerion concept machine being released in 1995 embodies a considerable amount of state of the art technology by Claas and associated partners in the project. With power outputs of 200 to 300hp and the new Claas HM8 stepless transmission giving a full speed range under load without speed range surge, it should become an ideal base unit for a wide number of attachments for large-scale farming.

It is shown below with a mounted Kleine sugar beet harvester undergoing trials. A 180° positional cab and four steering options are other interesting features.

The finger bar mower reigned supreme for a long, long time and was superseded only relatively recently by drum- and disc-type machines. One of the latest variations on the disc-mowing system, developed by J. Freudendahl, is the Hydro-Flex of swinging or articulated design. The operating mode of this machine can be likened to that of the reversible plough, i.e. working a field left to right. This gives an added bonus that will please conservationists, in that this method allows wildlife to escape via the unworked side whereas with the conventional round and round style, wildlife is trapped and silage can be polluted. Attachments are available to allow side-by-side swaths for subsequent pickup by 2.5 m foragers and swath on swath to allow for 1.8 m pickup.

Shelbourne Reynolds Engineering originated as the Bedford-based W. Reynolds company, whose products were straw trussers and pickup reels for many makes of combine. More recently they held the rights for Mather & Platt pea viners and vegetable harvesters, but for several years their stripper header attachments and swathers have been the principal products from this company now based in Suffolk. Initially many British farmers thought that the long stubble remaining after the abolition of straw burning might prove troublesome, but recent technology in terms of ploughing and tillage should overcome this problem. The John Deere 9600s illustrated are stripper harvesting rice at an average 8 to 10 acres per hour in California using SRE six-metre heads.

Facing page
Spalding, Lincolnshire has long been known as the UK equivalent of the Dutch bulb fields, so it is hardly surprising to find Dutch-built tulip toppers in action in that area.

AXIAL ROTOR

CONVENTIONAL

Crop types and climatic conditions can dictate the preference for either conventional or axial flow-type harvesters, and the possibility of exporting machines to countries with differing requirements should prove well within the capabilities of the FireFox 5000, which is shown here.

This innovative design, developed by the Rosso Corporation of Larizzate, Italy, can be adapted to either system within four hours and also offers full/half tracks with rubber-type option and two- or four-wheel drive. Rosso also manufacture track assemblies for combines and other agricultural applications.

In 1950 Minneapolis-Moline introduced what was probably the forerunner of the universal tool carrier or system tractor, particularly for handling a range of major equipment for harvesting applications. The Uni-Tractor base unit powered by their 48hp V206 motor was designed to operate with the Uni-Foragor, Picker Sheller, Huskor, Harvester or the Balemaster. Updating occurred in 1955 and continued under the ownership of New-Idea in the late 60s. A number of other major manufacturers followed suit with varying degrees of success.

This early example of the Balemaster 760, owned by MM enthusiast Ray Bailey of Wrexham, Clwyd, is not fully restored but is in full working order, as was demonstrated when this photo was taken in the 1990s.

Old ideas quite often have a habit of reappearing as new ones. In the late 1980s the Laverda MX series with down-front threshing cylinder (above) was available somewhat updated from the original concept in that the modular design allowed the machine to be readily broken down to the cutting table, threshing assembly and a compact base unit. Fiat-Agri/Laverda also cater for the hillside market with the AL models, one of which can be seen (right) undergoing tests.

85

Araus Hnos of Noetinger, Cordoba, Argentina are a family-owned company who manufacture a range of eight combine harvesters which begin with the compact model 280 and extend to the 590. As with certain other models in the range, the 280 has four-wheel drive and four-wheel steering which, with double planetary drive, make it a good machine for operating in less than ideal conditions. A basic or regular model is available as the 410, which has tanker or bagger options and a sun canopy in lieu of cab. Other range features include hydrostatic or mechanical drives, plastic reel fingers and pressurised or air-conditioned operator stations. Illustrated above is a recently completed model 505 seen at the factory.

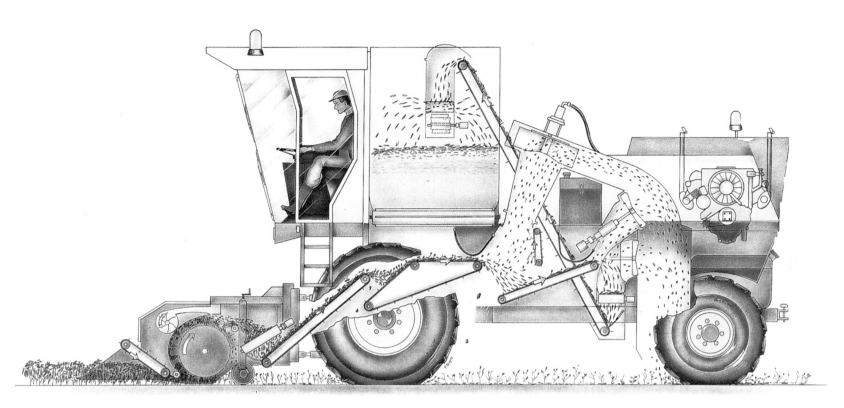

In addition to manufacturing an extensive range of harvesting machinery, Ploeger Machines of the Netherlands have had joint ventures with Riecam and the AVR company of Belgium. Harvesters for beans, peas, spinach and sugar beet are prominent products, and development was undertaken with hop harvesting, but owing to a change in brewery policy this project was not continued. Their BP 700 bean harvester is shown in schematic section.

J.I. Case and International Harvester have long been integrated into the Tenneco Corporation and currently trade as Case-IH. Case withdrew from combine production in 1972, but since the introduction of the first IH axial-flow rotary concept machines in 1977, over 80,000 units have been sold worldwide. Their long-running 1600 series was replaced recently by the 2100 range shown above. Features of the current machines include automatic reel speed, float control and return to cut and a unique single rotor concept. The histories of these companies go back a long way. Case built its 100,000th thresher and first combine in 1923. International Harvester dates back to Cyrus Hall McCormick's reaper in the 1830s. The International Harvester Company was formed in 1902, and the first combine from this group appeared in 1914 under the McCormick-Deering trade name.

While the All-Crop 60 set the standard for other major manufacturers to follow, as the final build total of over 250,000 units suggests, Allis-Chalmers were not afflicted by complacency. With greater acreages under cultivation, 1952 saw the introduction of the model 66 replacement, and this was followed by the 72 and 90 models. As with earlier machines, the Hart Scour Kleen weed seed removal system was available for harvesting seed crop. The auger feed replaced the canvas conveyor feed layout to give a greatly improved performance, and the Big Bin model soon became available. The 72 Big Bin illustrated at the right is one of three owned by the Lepper family of Hubbard, Iowa, who are blessed with 'orange blood' and report harvesting their soy bean crop at 3 mph at an average of 60 bushels per acre with no problems.

With over 200 units sold in the UK alone, the French Matrot company seem to have proved that six-row sugar beet harvesting is the way to go. The Deutz-powered M41 owned and operated by Maurice Edwards (pictured at the right) has been superseded by a model powered by a Mercedes diesel. With the option of 400hp, Matrot currently offer the most powerful sugar beet harvester in the UK.

Since the introduction of their first self-propelled forage harvester in 1972, the Claas company of Westphalia, Germany, have produced no fewer than 10,000 of this type of machine. Number 10,000 appeared in the UK in 1994, where it visited the Royal Smithfield Show in London, and it will be seen working in Europe during 1995, resplendent in its green and white metallic paint, walnut dashboard, chrome strips and '10,000 Jaguar' emblazoned under chairman Helmut Claas's signature.

The machine in question, a model 880, is seen at the front of the photograph, followed by an 860, 840 and 820 on demonstration. As Claas's combine harvester division has built well in excess of 300,000 machines, the company are certainly worthy of their 'Harvester Specialist' title.

Following the acquisition of the agricultural machinery assets of the Allis-Chalmers company in 1985 and the closure of the West Allis tractor plant and other facilities, the new owners Klöckner-Humboldt-Deutz of Cologne, Germany created considerable changes in American agriculture. One aspect of the new ownership that did not change was the harvester plant in Independence, Missouri.

In retaining this facility, KHD maintained the manufacture of rotary combines while the conventional range were phased out. They also used the plant as a base for handling the green, air-cooled Deutz tractors and the like from Germany.

The year 1991 saw the clock turn full circle with a senior management buy-out and the renaming of the firm as the Allis-Gleaner Company (AGCO). This transaction was followed soon after by the purchase of White, New-Idea and Hesston farm machinery interests, and more recently

the Massey Ferguson agricultural machinery rights were added.

The current range comprises the R42, 52, 62 and 72; the 330 bushel capacity of the 72 makes it the largest capacity machine in the United States at present. Deutz air-cooled power was standard, but with Cummins optional on the three smaller models and retaining the original A-C Natural Flow transverse cylinder.

Shown above is an R52 purchased in 1993 by Rodney Franzen from the Paul Kuhn dealership in Fort Atkinson, Iowa. It is ready for the summer oat harvest with a windrow pickup head.

The history of the noted F.A. Standen Company of Ely in East Anglia goes back almost 150 years. It shows a long involvement with sugar beet and potato harvesting machinery, which includes an association with Maschinenfabrik Niewöhner, a leading German company whose products bear the Wühlmaus name. The privately owned Standen Engineering Ltd have quite recently established an associate company known as Standen Reflex, which incorporate the Reflex Engineering company and now handle the products of Asa Lift, Herriay, O.Y. Juko, Baselier and the latest in high-capacity six-row beet harvester from Holmer, whose features include an on-board computer, co-driver facility, even-weight distribution regardless of hopper content and power by a 272kw/370hp MAN diesel. Illustrated to the left above is the latest Spectrum three- or four-row sugar beet harvester (shown at the Royal Smithfield Show). The Spectrum, operated in conjunction with a tractor front-linkage-mounted Turbo Topper, features auto depth / auto row seeking, and a hillside steerage kit is optional. To the left below is a compact one-row Juko semi-mounted potato harvester. On the facing page is the latest Holmer, shown in action with the defoliator in full flow.

Although self-propelled balers are not at all a new idea (Jones Balers Ltd produced their Invicta in the 1950s), machines of the size potential of the recently announced Deutz-Fahr Power Press 120H are certainly crowd-pullers. Based on Deutz combine componentry allied to a Fortschritt 550 baler mechanism, this 222hp, 8.4 m behemoth has a rear-mounted camera operating in conjunction with an in-cab video that gives the operator a view of the baling process. The four-bale accumulator can produce up to 110 bales per hour. As the machine carries a price tag of up to £144,000, it will be of chief interest to contractors and large farming groups.

Bell of South Africa produce a variety of equipment for handling sugar cane like the trailers and loaders shown here.
During the 1960s, the late Salopian Engineers Ltd made a small batch of handlers under licence, but apart from a single machine shipped to the West Indies for trials, the venture was not continued and the power units were removed and sold in Scandinavia.

95

In 1992 Mather & Platt (FMC) announced their latest green bean harvester, the model BH7100 from their French plant acquired during the 1980s. As France is the largest bean-growing country in Europe, this would seem to be a logical move by FMC. The machine illustrated is harvesting Odessa beans.

Header sizes of 30ft are usually associated with the Americas, Australia and the Middle East. However, the hedgerow removal policy of the past few decades has seen the emergence of much larger and more accessible fields on arable farms in the UK, and Claas UK Ltd recently introduced their first 30ft Commandor 228 into this country. The Humberside-based CWS Farms, with 3500 acres of combinable crops, expect their new flagship machine shown above to achieve wheat outputs of up to 30 tonnes per hour.

This Taarup forager mounted on a reverse-drive industrial 280hp Steyr tractor was an experimental project undertaken by WestMac Ltd but not pursued on a commercial basis owing to technical problems relating to the cab structure. The reverse-drive concept is not altogether new. At least one other development was undertaken with this in the early 1980s with the Massey Ferguson tractor-based combine harvester, which also did not become a commercial proposition.

Facing page
The project to develop a reverse-drive combine harvester involved design work by a private engineer and the Massey Ferguson works at Eschwege in Germany. The machines were offered in sizes 4.80m/16ft for mounting to tractors of 110–150hp and 5.60m/18ft for 120–200hp tractors, with crawler mounting for rice and maize harvesting operations being optional. The rights were later divested to the Mörtl company.

Following the appearance of the Mobil-Trac system on the Caterpillar Challenger tractor in 1989 (shown above), Claas adopted this innovative rubber track as the Terra-Trac system for their harvesting products. The basic principle of this design is to combine the advantages of both tracked and wheeled layouts for applications where reduced ground pressure is a priority. Shown below is a Commandor combine based on this system.

100

A long-term involvement in harvesting machinery is evident in the latest oil seed rape swather from SRE Ltd. Originally based on Fortschritt tractor units, the current Mentor series is based upon Canadian McDon Industries units of 96hp and two-speed hydrostatic transmission and total operator consideration. These units are matched to in-house custom-built header units that feature hydraulic functions, six bat flip-over anti-wrap reels and other features essential for harvesting this particular crop.

Environmental awareness together with the need to conserve the world's dwindling fuel resources has led to oil seed rape being assessed as an alternative biofuel. Should this prove to be a viable use, machines of this type will be very important in the future.

The TF and TX combines from New Holland feature what must be the ultimate in design and technology of operator stations. Their new Discovery cab shown above has, in addition to the in-built electronic monitoring and optimum visibility, a further advantage for the operator in that engine noise has been reduced to a very low level because of the way in which the grain tank has been installed.

As raspberries are very soft structured fruits which ripen individually on last year's canes while surrounded by soft new canes, they would appear to defy mechanical harvesting methods, but Pattenden Machinery have succeeded with help from organisations such as the Scottish Crop Research Institute and the Scottish Soft Fruit Growers Ltd, which administer the Raspberry Restructuring Scheme for the European Community.

The Osprey, shown below, can pick ripe berries for the individual quick frozen (IQF) or the pulp market to a standard that some say is better than hand picked. Features include accurate stroke and frequency control of the shaker, a ground speed sliding tray catching system and minimum fruit drops. Pattenden also feature the Challenger blackcurrant harvester, shown above.

The top of the range plot combine from Hans-Ulrich Hege, the 180, is shown here. It contains the componentry and self-cleaning characteristics of the smaller 140 and can handle all sizes of trial plots, experimental fields and yield trials, etc. Two-speed hydrostatic drive for field and on road operation, foot pedal stop-go facility and diff lock are some of its features. Other trial plot machinery from this specialist manufacturer includes the 16 laboratory thresher and the 212 green forage harvester, the latter being designed specifically for crops such as grasses, clover, alfalfa, forage maize, herbs and other types of forage. Power is provided by the Volkswagen Golf 31kw/42ps diesel engine.

As a result of the recent increase in environmental awareness, research is under way to find 'friendly' regenerative fuels for power stations to replace emission-prone fuels such as coal, etc. Trials have been undertaken with willow as such an alternative fuel, and Claas forage harvesters (right) are easily adaptable for this purpose. Another possibility is to convert sugar cane harvesters as shown in the illustration below.

Facing page
Some good news for vegetable harvester operators in the 1990s concerned the new John Deere 6000 series tractors with option of a creeper transmission. Cornishmen Tony and Peter Hendy purchased their first Deere, a 75hp model equipped with the nine forward/three reverse creeper option, to operate in conjunction with a Burdens Mini Veg Packer to follow the hand-pickers at speeds down to 220 metres per hour. At present this is the closest you can get to full mechanisation for harvesting cauliflowers.

Hemos of Meppel in the Netherlands obviously had the larger acreage farmer and/or custom operator in mind with their Multi-Trac introduced in 1992. It is basically a tractor with hydraulic power, four-wheel steering (front, rear, front and rear, and crab modes) and power from a lusty 420hp V8 diesel engine which can be located in different positions to suit various applications. The cab and pickup reel are from Claas, and the fully laden weight is 20 tonnes.

Facing page
Sampo Rosenlew Ltd, the well-known Finnish manufacturers of trial plot combines and one-time supplier to Massey Ferguson for their smaller models, have moved into the higher capacity market with their SR2045, 2050, 2055 (illustrated) and 2060 series. Features include cutting widths from 3.10 to 4.20 m, low ground pressure, rugged construction and optionals such as air conditioning and table header reverse.

Pull-type and mounted forage harvesters of small to medium size are now usually confined to smaller farm operations, whereas self-propelled models are of greater benefit to custom operators and machinery co-operatives where the greater output and manoeuvrability under adverse ground conditions can justify the higher price tag associated with these machines.

The 6000 series from John Deere in 1992 is representative of these high-capacity machines and come in four sizes up to 410hp with pickups of 3 and 4.5 m, four- and six- row maize headers and a Quick-Tatch mounting available for the four-row size. The 6710 is illustrated.

Byron Enterprises are a relatively young company dating back to the early 1970s but their range of corn and bean harvesting equipment embodies well-proven designs along with the latest technology. The 8400 Agmaster shown can handle fresh market corn or the sweet variety at up to 60 tph with its four-wheel hydrostatic drive.

Self-propelled potato harvesters are more widely operated on mainland Europe than in the UK. NV Dewulf, based in Belgium, specialise in self-propelled harvesters for potatoes, leeks, carrots, sugar beet, etc. Their model R3000S two-row potato harvester is seen working, stone and trash ejection being clearly evident.

Natural Technology manufacture this aquatic weed harvester which has won many innovation awards. It uses conventional conveying and hydraulics technology, squeezing out the water (which is 90% of the plant weight) before reducing plant volume in a compactor and ejecting it via a shredder/blower unit discharge tube. When the machine is on the water, the front elevator conveyors are moved forward and inclined with the 'jaws' at or below the water surface, wherever the weed mass is situated. Aquatic weed creates enormous problems in irrigation, flood defence and hydropower operations worldwide. One of NT's harvesters was purchased by the European Union for use on Lake Victoria, Uganda in 1994.

A Reed & Upton mobile pack house for ultra-large vegetable growers. It can accommodate up to 30 workers.

High tech is certainly the operative term for what is probably the most advanced development in harvesting technology today. The Massey Ferguson yield mapping system uses the Global Positioning System (GPS), whereby signals are transmitted 12,427 miles from a satellite to an aerial and electronic black box which converts the signals to give a location fix on the combine to an accuracy of within 5 m. The object of the operation is to provide the farmer with a yield map in order to assess future requirements regarding tillage, seed rate application and fertiliser/agrochemical dressing. The maps also help pinpoint areas of fields needing further investigation, such as for soil types, weed location and drainage efficiency, all of which are very important in commercial and environmental terms.

From the operator's point of view, little more is required than to press the relevant button at day's end in order to off-load the data from the combine's on-board computer onto a chip card which is then transferred to the farm computer so as to compile and print the yield map.

The current MF range as sold in the UK comprises the 23, 26, 30, 32 34, 36RS, 38 and the 40RS models. Features include Freeflow or Powerflow tables, automatic cutting heights,

auto self-levelling, and automated control of reel speed and forward speeds. Valmet diesels from 73.6 kw up to 215 kw power the range. From the MF30 and upwards, Datavision comprehensive monitoring is standard, with yield metering being optional. Worldwide, the MF combine range extends to 18 models and includes specialist plot models as well as baggers and rotaries. While the majority are built in Europe for home markets, markets in the southern hemisphere are mainly served by the company's USA-built rotary combines and a Brazilian-made tine has been developed particularly for rice, soya, corn and cereal harvesting.

As a result of the increased awareness of the need to develop renewable resources, Claas, in conjunction with several other companies and universities, have designed and built a machine to harvest miscanthus (elephant grass), a crop with a considerable growth rate and high biomass yield. The machine illustrated above combines forage harvesting and big baling techniques.

Acknowledgements

Agricultural Engineers Association
Agripress Publicity Ltd
Alexander & Duncan
E Allman Co
Maria Elena Araus
Manfred Baedeker
Ray Bailey
Gordon Bridge
Billy Brinksman
John Briscoe
Dr J H Brown
J Burley
J Calvera
C L & S Cawood
Brian Davis
DCN Associates Ltd
F J Eames
Maurice Edwards
Ron Eggen
John & Steve Emmett
Glyn & David Evans
Farm Contractor
Judy Farrow
Henry Flashman
Scott Freeman
M A Furness
Stuart Gibbard
John Gilbert
J L Llera Gil
Jim Golding
Steven Grimshaw
Harper Adams Agricultural College

Don Harris
Hans-Ulrich Hege
G. Hartnell
Peter G Houghton
International Rice Research Institute
Barry Job
James O Johnson
John & David Jolley
Siem Kamper
Ed Karg
John Kilgour
Professor Nickolai I Klenin
D & C Latham
Derek Lea
Arland Lepper
Roger Marshall
Bill McKenzie
John Melloy
Linda Meulemeester
Keith Miller
Steve Mitchell
Steven W Moate
Robert Moorhouse
Mike Neal
Newcastle-Under-Lyme Library
David Nicholson
Chris Nunn
Old Allis News
Ontario Agricultural Museum
Anna Patrick
Pharo Communications
Bob Powell

Processors & Growers Research Association
Graham E A Rand
Laurence Rooke
M.E.R.L,University of Reading
Richard & Sandra Schleichter
Calvin L Schmidt
Uwe Schreiber
Silsoe Engineering (formerly AFRC)
Peter Small
Brian Smith
Martin Smits
Dr Pavel P Sorokin
Staffordshire College of Agriculture
Ron Stanbrook
Des and Dilwyn Thomas
Joe L Thomas
Edward Thompson
Dick Tindall
Juan J Vallado
R Verschaeve
Jim Wallace
Howard Walsh
Tony Walters
Emerson Wertz
Peter Wheeler
Ken D Wickham
George Wisener
Jim Worden
Wye Valley Heritage Centre
Graham Yates
Dr Nikolay Zhiltsov

Enthusiasts' Magazines and Clubs

Australia

The Old Machinery Mart
Box 5237 MC
Toorsville
Queensland 4810
Australia

Germany

Fahr-Schlepper Freunde EV
Edelweiße 7
D-7702 Göttmadingen
Germany

Lanz Bulldog Club
Peter Petersen EV
Holstein
24211 Preetz
Germany

New Zealand

Country Life (NZ)
R H Robinson
Hamurana Rd
Nogongotaha RD2
Roturua, NZ 3221

United Kingdom

Friends of Ferguson Heritage
PO Box 62
Banner Lane
Coventry CV4 9GF
UK

National Vintage Tractor &
    Engine Club (Vaporising)
1 Hall Farm Cottage
Church Lane
Swarkestone
Derby DE7 1JB
UK

United States

Engineers and Engines
1118N Raynor Avenue
Joliet
Illinois 60435
USA

Gas Engine Magazine
PO Box 328
Lancaster
Pennsylvania 17603
USA

Old Allis News
10925 Love Road
Bellevue
Michigan 49021
USA

Massey Collectors News
PO Box 529
Denver
Colorado 50622
USA

Index

Page numbers in italics indicate an illustration. There may also be textual information on these pages.

A

Agmaster 8400, *111*
All-Crop 40, 19, *24*
All-Crop 60, 17, 18, *21*, 28, *46*,
    89
All-Crop 100, *24*
Allis-Chalmers 28, 30-1, 91,
    5000 series, 29
    All-Crop 40, 19, *24*
    All-Crop 60, 17, 18, *21*, 28,
        46, 89
    All-Crop 100, *24*
    forage harvester (1940), *19*
    Gleaner, *29*, 30-1, *44*, *45*, 91
        M2 model, *40*, 41
        MH hillside model, *41*
    Hart Scour Kleen weed seed
        removal system, 89
    Hurricane Junior, *28*
    R52, *91*
    Super A, *29*
Allis-Gleaner Company
    (AGCO), 91
apple harvesters, *47*, *48*
aquatic weed harvester, *112*
Araus Hnos model 505, *86*

B

Baldwin, Curtiss, 45
Bamfords of Uttoxeter Ltd., 31
bean harvesters, *70*, *87*, *96*, 111
Bell, the Rev. Patrick: cereal-
    reaping machine, *7*
Bell sugar cane equipment, *95*
Berry, G.S., 10
biofuel harvesters, *101*, *107*,
    115
Bisset binder, *59*
blackcurrant harvester, *104*
Bob 'Bean King' harvester, *70*
Bolinders-Munktell, 37
Braud grape harvester, *74*, *75*
brussels sprout harvester, *58*
Bulmer cider company, 47
Burley, John, 57
Byron Enterprises Agmaster, *111*

C

Carroll, Tom, 20

Case-IH 2100 range, *88*
Caterpillar, 8, 10
    Challenger tractor, *100*
cauliflower harvester, *106*, 107
Cebeco
    400 container chopper, *68*
    sugar beet harvester, *77*
Challenger blackcurrant
    harvester, *104*
Challenger tractor, *100*
Cicoria ATX 1600, *76*
Claas, August, 64
Claas, 8, *109*
    Commandor combine, *97*,
        100
    Crop Tiger, *67*
    flax harvester, *frontispiece*
    forage harvesters, *90*, *107*
    Hückepack, *28*
    Mähe-Dresch Binder, *16*, 17,
        64
    Matador, 27, *36*
    miscanthus harvester, *115*
    SF, *26*, 27
    Super series, 17, *20*
    Xerion concept machine, *78*
Claeys M80, *31*
Clayton and Shuttleworth
    'Combined Harvester and
        Thresher', *13*
combine harvesters (text), 6, 8-9
    Russian, 53-5
Commandor combine, *97*, *100*
Commander potato harvester,
    39
Communar combines, 53
Crop Tiger, *67*
Crusader, *37*

D

date harvester, *70*
Deere *see* John Deere
Depoortere flax puller, 60, *61*
Deutz-Fahr
    M66TS, *32*
    M660, 32, *33*
    Power Press 120H, *94*
    straw chopper/spreader, *69*
Discovery cab, *103*
Don 1500, 54, *55*

E

East European 'mystery
    machine', *46*
Edwards Vibro harvester, *47*
Emissey combine, *53*

F

Fahr, *19*, 31
    see also Deutz-Fahr
Fauré (inventor of mechanical
        pea sheller), 49
Ferguson, Harry, 27
Ferguson TE20, *27*
Fiat-Agri/Laverda
    AL model, *85*
    MX series, *85*
Firefox 5000, *82*, *83*
flax harvesters, *frontispiece*, 60,
    61
Fleming, Bob, 21
FMC, 50-2
    bean harvester, *96*
    Pea Picker, 51-2
        879 model, *51*
        979 model, *front cover*, 4
    Planetary Threshing System,
        50-1
    see also Mather and Platt
Ford, Henry, 27
Ford Motor Company, 37
Fordson model F, *44*, 45
Fortschritt
    combines, 35
    E175, *34*, 35
    see also Mähe-Dresch-Werke
Freudendahl, J. (JF)
    Hydro-Flex, *79*
    wrap-around combine, *45*

G

Gleaner combine, *29*, 30-1, *44*,
    45, 91
    M2 model, *40*, 41
    MH hillside model, *41*
Grain Marshall, *72*
grape harvester, *74*, *75*
Grimme
    Commander potato
        harvester, *39*

Continental harvester, *42*
Guaresi model 6-89, *71*

H

Haldrup
    1500, *60*
    trial plot swath-mower, *62*,
        63
Hart-Parr tractor, *12*
Hart Scour Kleen weed seed
    removal system, 89
'Harvest Brigade', 20
Hege
    122C plot thresher, *63*
    125 plot combine, *60*
    180 plot combine, *105*
Hemos Multi-Trac, *109*
Holmer sugar beet harvester,
    92, *93*
Holt, 10
    see also Caterpillar
horse-drawn machinery, 10, *12*,
    17
Hückepack, *28*
Hurricane Junior, *28*
Hussey, Obed, 7
Hydro-Flex, *79*

I

International Harvester, 53, 88
    apple harvester, *48*
    B64, *72*

J

Jamaffa brussels sprout
    harvester, *58*
JF *see* Freudendahl
John Deere, 25
    6000 series, *106*, 107, *110*
    9600, *80*
    model 55, *24*
    Titan II, *66*
Jolley family collection, *72*
Jones Balers Ltd., 30-1, 94
    Cruiser, *30*
    Pilot, *30*
Juko potato harvester, *92*

K

Kenworth truck, *66*
Kilgour, John, 57, 70
King Saud University, 70
Kleine sugar beet harvesters, *77, 78*
Klöckner-Humboldt-Deutz, 19, 91
Ködel & Böhm, 19
    thresher, *35*
Kolos combine, 53

L

Landmaschinen VEB, 35
Lanz
    Bulldog tractors, 11, 25
    MD2405, *25*
    steam engine and thresher, *11*
lavender harvesters, *back cover, 4*
Laverda
    AL model, *85*
    MX series, *85*
Lilleston bean harvester, *70*

M

Mähe-Dresch Binder, 16, 17, *64*
Mähe-Dresch-Werke, 35, 60
maize header, *76*
Maschinenfabrik Niewöhner, 92
Massey Ferguson, 91
    Mörtl reverse-drive combine, 98, *99*
    Power Flow table, 42, *43*, 114
    satellite range combine, *114*
    tractor-based combine harvester, 98
    tractors, *38, 45*
    yield mapping system, *114*, 115
Massey Harris, 8, 12, 27
    726, *22*, 23, *72*
    735, 27, *72*
    Clipper, 23
    hillside Gleaner, *41*
    peanut picker, *17*
    potato harvester, *17*
Massey Manufacturing Company, *11*
Matador, 27, *36*
Mather and Platt, 50, 80
    BH7100 bean harvester, *96*

Matrot M41 sugar beet harvester, *89*
McCormick, Cyrus Hall:
    original reaper, *7*, 8, 88
McDon Industries Windrower, *back cover*, 4
McKay Harvester Company:
    Sunshine harvester, *12*, 14
Mercer, R.F. & M.: Jamaffa brussels sprout harvester, *58*
Merritt, H.C., 18
Michigan combine, 10
Minneapolis-Moline:
    Balemaster 760, *84*
miscanthus harvester, *115*
Mitchell Cotts tea combine, *56, 57*
Moore & Hascall: Michigan combine, 10
Mörtl reverse-drive combine, 98, *99*

N

National Institute of Agricultural Engineering, 24
Natural Technology: aquatic weed harvester, *112*
New Holland, 31
    combines, *65, 102, 103*
    Discovery cab, *103*
Niva combine, 53
Norfolk Lavender Ltd., *back cover*, 4
Northern model combine, 53
NV Dewulf R3000S, *111*

O

oilseed rape harvesters, *63, 101*
Oppel Inc., 28
Osprey raspberry harvester, *104*

P

Paccar Company, 66
Pattenden Machinery
    Challenger, *104*
    Osprey, *104*
pea harvesters (text), 49-52
    FMC Pea Pickers, 51-2
    879, *51*
    979, *front cover*, 4
    pull-type viner, *50*
    swather, *50*

peanut picker, *17*
Pearson, Richard, 39
    Commander potato harvester, *39*
    Grimme Continental, *42*
Ploeger Machines: BP700 bean harvester, *87*
potato harvesters, *17*, 28, *38, 39, 73, 92, 111*

R

Ransomes, Sims & Jefferies Ltd.
    Crusader, *37*
    portable steam engine and stationary thresher, *9*
raspberry harvester, *104*
Reed & Upton mobile pack house, *113*
rice harvesting, 52, 54, 67, 80, 98, 115
Root Harvesters Ltd: Super Duplex potato harvester, 72, *73*
Rosso Corporation: Firefox 5000, *82, 83*
Rostelmach combine plant, 53, 54
Rumely No. 3, 18
Russian combines, 35, 53-5

S

S1, S4 and S6 Russian combines, 53
Salopian Engineers Ltd., 95
Sampo Rosenlew combines, *108*, 109
Samuelson, Bernhard, 8
Samuelson Co. Ltd: self-raking reaper, *8*
Schandoney hitch, *12*
Shearer cut header harvester, *15*
Shelbourne Reynolds Engineering (SRE), 80
    Mentor oilseed rape swathers, *101*
    stripper header, *65*
Sibiryak combine, 53
Silsoe Research Institute vegetable gantry, *74*
SK3 Russian combine, 53
SK10 Russian combine, *54*
Spectrum sugar beet harvester, *92*
Sperry New Holland, 31
    see also New Holland

SRE *see* Shelbourne Reynolds Engineering
Standen Company, 92
    Spectrum sugar beet harvester, *92*
Steyr tractor, *98*
straw, 8-9, 80
    chopper/spreaders, 53, *69*
sugar beet harvesters, 28, 71, *77, 78, 89, 92, 93*
sugar cane harvesters, 28, *95, 107*
Sunshine harvester, 12, 14
Super Duplex potato harvester, *73*

T

Taarup forager, *98*
tea harvesting, 56-7
Tenneco Corporation, 88
Thompson Company, 28
Titan II combine, *66*
tomato harvester, *71*
Toro & Mityana Tea Company, 57
Toronto light binder, *11*
tulip toppers, 80, *81*
Turbo Topper, 92
Tuthill-Temperley fruit harvesters, 47, *48*

U

University of Agriculture and Forestry, Ho Chi Minh City:
    axial-flow thresher, *52*

V

vallus, 14, 65
vegetable gantry, *74*
Vicon potato harvester, *38*
viners (pea) *see* pea harvesters
Volvo, 31, 37

W

Westmac Ltd., 98
Wild, M.B. 'Harvester Thresher', *23*
Williams tea harvester, *57*
willow harvester, *107*
Windrower, *back cover*, 4

X

Xerion concept machine, *78*

Y

yield mapping system, *114*, 115

FARMING PRESS BOOKS & VIDEOS

Below is a sample of the wide range of agricultural and veterinary books and videos we publish.
For more information or for a free illustrated catalogue of all our publications please contact:

Farming Press Books & Videos
Wharfedale Road, Ipswich IP1 4LG, United Kingdom
Telephone (01473) 241122   Fax (01473) 240501

Early Years on the Tractor Seat   *Arthur Battelle*

Starting in 1938, this is a first-hand account of how a lifetime's involvement with tractors got started. Humorous, detailed, enjoyable for enthusiasts and general readers alike.

Classic Farm Machinery   *VIDEO*
Brian Bell, narrated by Chris Opperman

A wonderful collection of archive film extracts from the period 1940–70 showing every farm activity from ploughing to harvest.

Fifty Years of Farm Machinery   *Brian Bell*

Over four hundred illustrations demonstrate the development of all kinds of farm machinery 'from starting handle to microchip', made by both famous and the less well known manufacturers.

Tractors at Work   *Stuart Gibbard*

A fascinating collection of photographs — many never before reproduced — showing the working history of tractors in Britain from the turn of the century until the present day.

Tractors Since 1889   *Michael Williams*

An overview of the main developments in farm tractors from their stationary steam engine origins to the potential for satellite navigation. Illustrated with over 150 photographs, a third in colour.

Ford & Fordson Tractors
Massey-Ferguson Tractors
John Deere Two-Cylinder Tractors
          Michael Williams

Three heavily illustrated guides to the major models.

Fordson, the Story of a Tractor   *VIDEOS*
The Massey-Ferguson Tractor Story
John Deere Two-Cylinder Tractors
(Volumes 1 & 2)

Videos showing the history and development of the machines produced by these leading companies.

Farming Press Books & Videos is a division of Miller Freeman Professional Ltd
which provides a wide range of media services in agriculture and allied businesses.
Among the magazines published by the group are
Arable Farming, Dairy Farmer, Farming News, Pig Farming and **What's New in Farming.**
For a specimen copy of any of these please contact the address above.